YOUR KNOWLEDGE HAS VALUE

- We will publish your bachelor's and master's thesis, essays and papers

- Your own eBook and book - sold worldwide in all relevant shops

- Earn money with each sale

Upload your text at www.GRIN.com
and publish for free

Bibliographic information published by the German National Library:

The German National Library lists this publication in the National Bibliography; detailed bibliographic data are available on the Internet at http://dnb.dnb.de .

This book is copyright material and must not be copied, reproduced, transferred, distributed, leased, licensed or publicly performed or used in any way except as specifically permitted in writing by the publishers, as allowed under the terms and conditions under which it was purchased or as strictly permitted by applicable copyright law. Any unauthorized distribution or use of this text may be a direct infringement of the author s and publisher s rights and those responsible may be liable in law accordingly.

Imprint:

Copyright © 2018 GRIN Verlag
Print and binding: Books on Demand GmbH, Norderstedt Germany
ISBN: 9783668672307

This book at GRIN:

https://www.grin.com/document/418219

N'kosi Craigwell-Walkes

Indicators of social and economic Development

GRIN Verlag

GRIN - Your knowledge has value

Since its foundation in 1998, GRIN has specialized in publishing academic texts by students, college teachers and other academics as e-book and printed book. The website www.grin.com is an ideal platform for presenting term papers, final papers, scientific essays, dissertations and specialist books.

Visit us on the internet:

http://www.grin.com/

http://www.facebook.com/grincom

http://www.twitter.com/grin_com

Name: N'kosi Craigwell Walkes

GEOG 1132: World Economy Agriculture and Food

Title: Indicator of Development

Development is defined as the standard of living of people economic development is supportive and it involves increased per capita income and creation of new opportunities in education, healthcare, employment sectors (Drewnewski, 1966). In the 1950s and 1960s, development was mainly looked at through an economic lens and a country was developed was based on the standard and output of a country's economy. A more overall view began to take place in the 1970s as aspects such as poverty, health and education started to be considered and recognised as social issues that resulted from trying to achieve economic development. This lead to the birth of the measurement of social development and the emergence of social indicators of development. The two categories of economic and social indicators of development facilitate a more wholesome way of analysing and determining development. Each of these indicators has its own importance that helps to classify countries development and their economies which is what this essay will be looking to discuss.

As said before indicators have been separated into both economic and social indicators. Economic development is a concept that has many dimensions as there is no single measure of development that completely captures the process. Economic development is generally defined to include improvements in material welfare especially for persons with the lowest incomes, the eradication of mass poverty with its correlates of illiteracy, disease and early death, changes in the composition of inputs and output that generally include shifts in the underlying structure of production away from agricultural towards industrial activities, the organization of the economy in such a way that productive employment is general among working age population rather than the situation of a privileged minority, and the correspondingly greater participation of broad based groups in making decision about the direction, economic and otherwise, in which they should move their welfare (Herrick 1958). There are three very important economic indicators that aid in measuring a country's development. They are Gross Domestic Product (GDP), Gross National Product (GNP) and Purchasing Power Parity (PPP). To measure national economic development, to make an assessment of the economic performance of a country as well as for measuring standard of living of the people per capita, the Gross domestic product which is also known as per capita income is generally used as an important indication for the observation and monitoring of economic growth trends. The gross domestic product is the total value of goods and services produced by a country in a year (BBC, 2014). Most of the economic planners and forecasters have used GDP per capita because it helps to acknowledge economic welfare. It plays a key role in developing policies and plans for development due to the fact that GDP per capita illustrates whether an economy is progressing or not in a more understandable manner. It assists in creating a benchmark or a standard for both policy-making and comparisons between countries' economies.

The gross domestic product is also used as it is very comprehensible so much so that it is included in the human development index which is a social indicator of development. This indicator is also seen an held as a replacement for all the other economic activity because an increase in the national income can be seen as the most relevant, convenient, single measurement of development for both poor and rich countries (Baldwin, 1976). Most countries that are under the United Nations membership provide annual estimates of their gross domestic product per capita and this enables the ability of comparisons between countries' economy to be less difficult and have greater substance. Gross domestic product per capita al focuses on taking the mantle away from the gross national product as it gives a better indication of poverty and standard of living thus allowing measures to be put in place to help in the improvement of the poverty line and the standard of living of the population. Gross domestic product per capita income can also be used to classify a country status in two categories known as forward and backward. Countries and regions that are classified as backward usually associate with high birth rate high death rate, small amounts of saving and low level of agricultural productivity while forward countries are the opposite of this with low birth rates, high life expectancy and high amounts of saving. The gross domestic product is also an important indicator for measuring economic production of a country such as saving, credit and accumulation of capital (Allan County Community College, 2016).

Another indicator that is generally associated with the gross domestic product is Gross National Product (GNP). The gross national product is an estimate of total value of all the final products and services produced in a given period by the mean of production owned by a country's resident (Investopedia, 2018). It is generally calculated by taking the sum of personal consumption expenditures, private domestic investment, government expenditure, net exports as well as income that is earned by residents from overseas investment and the subtraction of the income earned within the domestic economy by foreign residents. Gross national product measures the total monetary values output that is produced by a country's residents. This means that the output that is produced by any foreign residents such as Tran-National Companies or tourists in the country's boarders has to be excluded in the calculations and any output produced by residents of the country that live outside of the boarders should be acknowledged.

Most times both gross domestic product and gross national product are consistently linked together and confused mainly because they have related concepts. The most significant and identifiable difference and overlapping that sometimes cause confusion between the two is that there are sometimes companies owned by foreign residents that produce goods within a country and companies owned by the person that is from and live within said country, however, produce products that exported throughout the world and revert earned income to domestic residents. (Investopedia, 2017). Such example of this could be observed in a country such as the United States where there is a number of foreign companies that provide product and services within the United States and then transfer any of the income earned to their foreign residents. The same could be seen by numerous United States companies that produce the goods and services outside of the United States, however earning their profits for United States residents. (Investopedia, 2017).

Both the Gross Domestic Product and the Gross national product has there limitation or criticisms and most of them are similar for both indicators. Countries with a high GDP per capita are sometimes countries that have a low educated population or a low satisfactory educated population (Allan County Community College, 2016). There is a report that the correlation between per capita and levels of education are not always positive as well as that a high GDP does not always mean greater welfare or production. This can be seen by a 2010 report between the years of school and a countries GDP per capita. The United Arab Emirates had a gross domestic product of approximately $44,000 USD but their average schooling years was only an estimated 9 years. This was extremely contrasted by the country of Czech Republic which had a gross domestic product per capita of approximately $20,000 USD but an average schooling year of an estimated 13 years (Alfonso, 2010).

Another limitation of these indicators is the economic cost versus the social costs. There is always the same problem that arises when analysing the economic and social implications that arise because there is no identity between the economic cost of producing the current national output and social the problems of the output (Allan County Community College, 2016). The economic cost is that include the variables such cost, indirect business taxes, capital consumption and allowance for which some kind of monetary values is possible while the social costs related to intangible factors such as the general destruction of the physical, natural and social environments as a result of current production (Agarwal, 2016). An example of is the beauty of the country-side may be destroyed beyond repair mostly happens in countries where there is heavy mining and is highly industrialised countries. Rivers and the atmosphere sometimes get polluted during the removal and disposal of natural wastes, diseases. There is also the emergence of social negative such as crime that could arise. It is difficult for indicators such as gross domestic product and gross national product to highlight and acknowledge these aspects that comes with development.

Lastly, gross national product and the gross domestic product does not take into consideration the distribution and inequality of the country's wealth and between urban and rural areas. Since economic continues to progress, inequality in the distribution continues to be a big problem as without its reduction there would not be any economic development. When a small amount of percentage of the gross domestic product is owned by only the wealth of the society then there is little to no economic development in the majority of the country's economy. This is a problem in the country of Saudi Arabia where the wealth of the country is not shared equally and is mainly shared amount the elite of the society (Allan County Community College, 2016). The United States is also a country where inequality and uneven distribution of wealth is a growing problem. The United States has 41.6 percent of the global personal wealth and yet even with this overall growth of wealth, there is still a reoccurring issue of inequality (Fortune, 2018). The Allianz calculated each country's wealth Gini Coefficient which is a measure of inequality in where 0 is perfect equality and 100 is perfect inequality and it found that the United States had the highest most wealth inequalities with a score of 80.56 which shows the concentration of overall wealth in the hands of the proportionately fewest people (Fortune, 2018).

Another economic indicator is Purchasing Power Parity (PPP). Purchasing power parity is the rate at which the currency of one country would have to buy the same amount of goods and services in each country (Callen, 2017). The concept associated with purchasing power parity could only be fully understood by one learning about the Big Mac theory. The Big Mac Theory was developed by economist George Ritzer, who is an American sociologist that also studied globalization and patterns of consumption (University of Maryland, 2013). In his book 'The McDonaldization of Society' he talked about the Big Mac theory to explain the purchasing power parity. The Big Mac Index was the amount of time an average worker in a given country would have to work to gain enough money so that they would be able to buy a standard product with the product being the Big Mac. For example, a hamburger is selling in London for 2 pounds and the price for it in New York is 4 US dollars, it would allow one to come to the assumption that is one pound to 2 U.S dollars (Callen, 2017). Purchasing power parity exchange rates are relatively rarely ever fluctuation as compared to the market rates which are more erratic and ever-changing. Unlike market-based rates, purchasing power parity could be used to compare the power of the currency of countries for even non traded goods such as haircuts, taxi fares and tickets for matches (Callen, 2017). It also has to do with non-traded goods and services generally happens to be cheaper in low-income countries than high-income countries. For example, the taxi fare for the same distance is higher in Paris which is a high-income country than those in a low-income country of Tunis (Callen, 2017). This is sometimes due to the fact that countries with low income most of the services provided would be more commonly labour intensive compared to a high-income country and even though the tradeable goods such as machinery is the same price in both countries it would most likely be less to get for example a haircut in Lina than in New York. Purchasing power parity is also seen as a more significant measure of overall well-being as it does not lover look taking into account the differences in cost of non-traded goods across countries that could result in an underestimation in the power of consumers in newly developing markets and countries as well as there over well-being. (Callen, 2017).

Though purchasing power parity has many advantages to it, still has its drawbacks similar to other indicators. One of the biggest problems with the purchasing power parity is that it is harder to calculate and measure than market based rates. This is due to the fact that the purchasing power parity has to be estimated which gives room for inaccuracies being allowed into the measurement. There is also the effect that tariffs and transport cost can fluctuate and lead to a deviation of the short-term equilibrium exchange rate from the purchasing power parity, the amount of this deviation varying directly with the severity of imperfections (Anand, 2016). Lastly, another limitation of purchasing power parity is that a difficulty with still occur even if the entire population of commodities are used to develop the price measure in the country. The value of the parity would most times rely upon or be affected by the price level selected (Anand, 2016). The purchase power parity will most times vary with the weighting pattern of the price measure.

There are not only economic indicators that are used to measure development. There are social indicators such as life expectancy, infant and child mortality access to improved water and sanitation and access to adequate health care that is used to measure the development of countries however the major social indicator that is used to measure development and that is the Human Development Index. After years of domination of economic indicators such as the gross domestic product being used for the measuring of development in 1990, the United Nations Development Report decided to create the Human Development Index. The aim of it was to correctly observe development in much more than the small lens of the expansion of income and wealth (Najom, 1998). It comprised of health, education, the standard of living, human rights, political freedom and self-respect (Ray,

2008). In that same year of 1990, the human development report forms a composite index known as the Human Development Index. The human development was formed on the foundation of three core dimension of human development. The first was to lead a long and healthy life, to acquire knowledge and to have access to resources needed for a decent standard of living (Aziz, 2015).

The human development index comprises of four variables that are used in representing the three dimensions previously stated. These are life expectancy at birth which correlates the dimension of a long, healthy life, adult literacy rate and combined enrolment rate at the primary secondary and tertiary levels which is linked to the knowledge dimension and lastly the real gross domestic product per capita (at purchasing power parity (PPP)) which associated back to the dimension of resources needed for a decent standard of living (Aziz, 2015). The human development index uses a ranking system between the numbers of 0 and 1. The closer it is to 1 the higher its human development index would be. The most significant and remarkable component of the human development index is that it looks to combine both the economic and social aspects of measuring development to systematically and comprehensively analyse the comparative status of socioeconomic development for both developing and developed countries. The human development index is considered to be better than most other social development indices because it effectively facilitates the evaluation of the progress of countries resulting in inter-country comparisons and intertemporal comparisons of the living level (Gounder, 1994). Some of the countries with the highest human development rankings are Norway with 0.949 at number one, Australia and Switzerland with 0.939 at number two and Germany at number 3 with a value of 0.926 (United Nations, 2015).

Human development index also has its limitation and negative as other development indices. One of them is that the human development index is not a comprehensive manner of human development (Human Development Report, 2016). This means that it only really pays attention to the basic development of human development and does not take into account several other dimensions of human development. Secondly, the index is composed of more long-term human development which means that it does not take into account newly implied policies as well as measure recent human development achievements. Since the human development index has other development indices such as life expectancy and gross domestic product, this means that the HDI also shares the limitation and negatives of these indices as well. Lastly, since the human development index is only as average this means that there is are many disparities and inequalities within countries. There is disregard of gender, regions, races and ethnic groups which can unmask the human development index and can be and has been used widely for policy formulation (United Nations, 2016).

As we could see there are several indicators for development. These differ in terms of which areas of development the focus on and well the variable they use to measure development. Even though they are separated into two categories of economic and social indicators which when all their data is grouped it helps to give a wholesome view to a country's or region's development and economies. However, though they have all these positives each indicator isn't perfect and have their own limitations and negatives about them that could be criticised leaves some flaws in each these indicators measurement of development.

References

Anon, 2014. *GCSE Bitesize: Economic development indicators*, BBC, from http://www.bbc.co.uk/schools/gcsebitesize/geography/development/contrasts_development_rev3.shtml .

Kästle, 2018. *The Human Development Index 2016*. Human Development Index 2016 - HDI - Nations Online Project, from http://www.nationsonline.org/oneworld/human_development.htm.

Anon, 2018. *Human Development Reports*. The Human Development Index – Human Development Reports, from, http://hdr.undp.org/en/hdi-what-it-is.

Anand, 2015. *Criticisms of Purchasing Power Parity*. Economics Discussion, from, http://www.economicsdiscussion.net/balance-of-payment/criticisms-of-purchasing-power-parity/6645.

Callen, 2017. *Finance & Development*. Finance & Development | F&D, from, http://www.imf.org/external/pubs/ft/fandd/basics/ppp.htm.

Afonso, 2015. *Education and GDP per capita*, from, https://alexandreafonso.me/2015/10/06/education-and-gdp-per-capita/.

Staff, I., 2016. *Gross National Product - GNP*. Investopedia, from, https://www.investopedia.com/terms/g/gnp.asp.

Fox, 1974. Combining Economic and Noneconomic Objectives in Development Planning: Problems of Concept and Measurement. *Economic Development and Planning*, pp.104–141.

YOUR KNOWLEDGE HAS VALUE

- We will publish your bachelor's and master's thesis, essays and papers

- Your own eBook and book - sold worldwide in all relevant shops

- Earn money with each sale

Upload your text at www.GRIN.com
and publish for free